# YOUR KNOWLEDGE HAS VALUE

**Bradley Tice**

# A Comparison of Compression Values of Binary and Ternary Based Systems

GRIN Verlag

**Bibliografische Information der Deutschen Nationalbibliothek:**

Die Deutsche Bibliothek verzeichnet diese Publikation in der Deutschen National-
bibliografie; detaillierte bibliografische Daten sind im Internet über http://dnb.d-
nb.de/ abrufbar.

**Imprint:**

Copyright © 2012 GRIN Verlag GmbH
Druck und Bindung: Books on Demand GmbH, Norderstedt Germany
ISBN: 978-3-656-64523-8

**This book at GRIN:**

http://www.grin.com/en/e-book/199142/a-comparison-of-compression-values-of-
binary-and-ternary-based-systems

**GRIN - Your knowledge has value**

Der GRIN Verlag publiziert seit 1998 wissenschaftliche Arbeiten von Studenten, Hochschullehrern und anderen Akademikern als eBook und gedrucktes Buch. Die Verlagswebsite www.grin.com ist die ideale Plattform zur Veröffentlichung von Hausarbeiten, Abschlussarbeiten, wissenschaftlichen Aufsätzen, Dissertationen und Fachbüchern.

**Visit us on the internet:**

http://www.grin.com/

http://www.facebook.com/grincom

http://www.twitter.com/grin_com

# A Comparison of Compression Values of Binary and Ternary Based Systems*

By Bradley S. Tice

*Abstract-* The paper will introduce the ternary, or radix 3, based system for use as a fundamental standard beyond the traditional binary, or radix 2, based system in use today. A compression level is noted that is greater than the known Martin-Lof standard of randomness in both binary and ternary sequential strings.

*Index Terms-* Radix 2, Binary, Radix 3, Ternary, Information Theory, Compression Ratio

## I. Introduction

A ternary, or radix 3, based system is defined as three separate characters, or symbols, that have no semantic meaning apart from not representing the other characters. This is the same notion Shannon gave to the binary based system used in his paper's on information theory upon it's publication in 1948 (Shannon, 1948). Richards has noted that the radix 3 based system as the most efficient base, more so than even the radix 2 or radix 4 based systems (Richards, 1955: 8-9). A compression level is noted in this paper that is greater than the known Martin-Lof standard of randomness in both binary and ternary sequential strings.

## II. Randomness

The earliest definition for randomness in a string of 1's and 0's was defined by von Mises, but it was Martin-Lof's paper of 1966 that gave a measure to randomness by the *patterlessness* of a sequence of 1's and 0's in a string that could be used to define a random binary sequence in a string (Martin-Lof, 1966). This is the classical measure for Kolmogorov complexity, also known as Algorithmic Information Theory, of the randomness of a sequence found in a binary string (Kotz and Johnson, 1982: 39). Martin-Lof (1966) also defined a random binary sequential string as being unable to compress from its original state. Nonrandom binary sequential strings can compress to less than there original state (Martin-Lof, 1966).

* Paper accepted and prepared for poster session Wednesday September 3, 2008 for the Royal Statistical Society 2008 Conference in Nottingham, England, United Kingdom, September 1-5, 2008.

The author contact information: Advanced Human Design, P.O. Box 3868 Turlock, California 95381 U.S.A., e-mail: paulatice@bigvalley.net

## III. Compression Program

The compression program to be used has been termed the *Modified Symbolic Space Multiplier Program* as it simply notes the first character in a line of characters in a binary sequence of a string and subgroups them into common or like groups of similar characters, all 1's grouped with 1's and all 0's grouped with 0's, in that string and is assigned a single character notation that represents the number found in that sub-group, so that it can be reduced, compressed, and decompressed, expanded, back to it's original length and form. An underlined 1 or 0 is usually used to note the notation symbol for the placement and character type in previous applications of this program. An italicized character will be used for this paper.

## IV. Binary System

The binary system, also known as a radix 2 based system, is composed of two characters, usually a 0 and a 1, that have no semantic properties except not representing the other. Group A will represent a nonrandom sequential binary string and Group B will represent a random sequential binary string. Both Group A and Group B will be 15 characters in total length.

Group A: [000111000111000] (Nonrandom)

Group B: [001110110011100] (Random)

Utilizing the Modified Symbolic Space Multiplier Program to process like sequential characters, either 0's or 1's, into sub-groups and note them with an italicized character specific to that sub-group and having it represent a specific multiple of that sub-group as found in a key, in this case Group A Key and Group B Key, as a compressed aspect to both Group A and Group B sequential binary strings.

Group A Key: All italicized characters will represent a multiple of 3.

Group B Key: The italicized character 0 will represent a multiple of 2 and the italicized character 1 will represent a multiple of 3.

Group A: [*01010*]

Group B: [*01011*010]

The compressed state of Group A, nonrandom, is five characters in length. The compressed state of Group B, random, is 8 characters in length. Note that the random sequential binary string in Group B compressed to less than the original total pre-compression length. This differs from standards known in Martin-Lof randomness and those found in Kolmogorov Complexity.


## V. Ternary System

A ternary, or radix 3, based system there are three characters used that have no semantic meaning except not representing the other two characters. Group C will represent a nonrandom ternary sequential string and Group D will represent a random ternary sequential string. The total length for each group, Group C and Group D, will be 12 characters in length. The three characters to be used in this study are a 0, 1, and 2.

Group C: [001122001122]

Group D: [001222011222]

Again each group will be assigned a specific compression multiple based on a specific character type, in this case an italicized 0, 1, and a 2, as defined in a key, Group C Key and Group D Key.

Group C Key: The italicized characters 0, 1 and 2 will represent each a multiple of 2.

Group D Key: The italicized character 0 will represent a multiple of 2. The italicized character 1 will represent a multiple of 2 and the italicized character 2 will represent a multiple of 3.

Group C: [*012012*]

Group D: [*012*0*12*]

The compressed state of group C, nonrandom, is 6 characters in length. The compressed string of Group D, random, is 6 characters in length. Again note that Group D, the random sequential ternary string, is less than it's pre-compressed state, and again, is novel for those extrapolations of binary examples found in Kolmogorov Complexity.

## VI. Application of Theory

Compression of data for transmission and storage of information is the most practical application of a binary system in telecommunications and computing. The application of a ternary, or radix 3, based system to existing communication systems has many advantages. The first is the greater amount of compression from this base, as opposed to the standard binary based system in use today, of random strings of data, and secondly, as a more utilizable system because of the three character, or symbol, based system that provides for more variety to develop information applications. From telecommunications to computing, the ternary based system applied at a fundamental standard would allow for a more robust communications system than is currently used today.

## References

Shannon, C.E. (1948)A mathematical theory of communication, *Bell Sys. Tech. Jour.*,27, 379-423 & 623-656.

Richards, R.K. (1955)*Arithmetic Operations in Digital Computers*, New York: Van Nostrand Company, Inc.

Martin-Lof, P. (1966)The definition of random sequences, *Info. & Cont.*, 9, 602-619.

Kotz, S. and Johnson, N.L. (1982)*Encyclopedia of Statistical Sciences*, New York: John Wiley & Sons.